SCIENCE

GRADE 2

WORKBOOK FOR KIDS

AGES 7-8

THIS BOOK BELONGS TO:

ISBN: 978-1-953149-89-3

Table of Contents

Unlock Your FREE Science Bonuses!

+ Answer Key

- **21 Hands-On Experiments:** Bring science to life with exciting, step-by-step activities you can do at home!
- **Learn Through Play:** Explore core concepts like Matter, Forces, and Energy with fun stories and characters for each experiment.
- **Sparks Curiosity:** A perfect companion to your workbook, designed to make learning interactive, messy, and memorable.

QR Code in the Back of the Book

Enjoying this Book?

We'd love to hear your thoughts

We may just send you something special.

Scan Me!

Introduction to Science Skills

Hey Brilliant Scientists,

Welcome to your next big science adventure! This isn't just any workbook, it's your guide to exploring even more amazing discoveries about the world around you. Get ready to investigate plants, animals, weather, energy, matter, and even how people use science to solve real problems!

Inside, you'll dive into fun stories, hands-on activities, exciting experiments, and awesome quizzes that will help you think, question, and explore like a true scientist. You'll observe animal habitats, explore Earth's resources, measure forces and motion, study the water cycle, and even learn how technology helps us in everyday life.

Grown-ups, you're part of the journey too! Talk about the stories, try the experiments together, and ask "why" and "how" questions along the way. Science becomes even more meaningful when it's shared!

Every activity in this book is designed to match what 2nd graders need to know and practice, building important science skills while making learning fun and unforgettable. With colorful activities, real-world connections, and lots of opportunities to experiment and imagine, this workbook is the perfect companion for curious, growing minds.

Thanks for choosing this science adventure. Now turn the page, grab your science tools, and get ready to explore, discover, and dream big!

Your friends at Polymath Panda

Unit 1 The Case of the Mysterious Objects!

Activity 1 The Case of the Mysterious Objects!

One sunny afternoon, Emma, the Science Spy, was walking in the park with her trusty magnifying glass. Suddenly, she saw a box sitting under a big tree. A note was taped to the box. It read:

"Dear Science Spy,
Inside this box are mysterious objects.
Can you figure out what they are? Use your best science skills! - The Secret Keeper"

Emma's eyes lit up. "A new mystery to solve!" she said. She opened the box and found all sorts of strange things: a smooth rock, a feather, a shiny button, a crumpled leaf, and a small seashell. "Time to use my science skills!" Emma said.

First, she used observation. She looked at each object carefully. She touched them to feel their texture. She noticed the rock was hard and cold, the feather was soft, and the button sparkled in the sunlight.

Next, she started classifying the objects. "How can I sort these?" she wondered. Then she got an idea!

"I'll group them by texture!" Emma said. She put the smooth rock and shiny button in one pile. She put the crumpled leaf and soft feather in another pile.

But she wasn't done yet. "What about size?" she asked herself. She noticed the feather and leaf were bigger, so she grouped them together. The rock, shell, and button were smaller, so they became another group.

By the time Emma finished, she had neatly sorted all the objects in two different ways. Just then, a new note fell out of the box. It said:

"Great job, Science Spy!
You used observation and classification to solve the case.
Keep exploring the world around you!" - The Secret Keeper

Emma smiled. "Science is the best!" she said as she packed up her magnifying glass and headed home, ready for her next adventure.

Object Detectives!

2 Activity

You have been recruited as a Science Spy on a top-secret mission!
Your task? Use your keen observation skills to find and classify mysterious objects based on their properties, just like a real scientist!
Get ready to explore, examine, and report your findings!

Instructions

1. Search for objects in your classroom, home, or outdoors that match the given descriptions.
2. Draw (✏), label (✍), and describe (✍) each object in the correct box.

Mission: Object Classification

Spy Challenge	Draw Your Object!	What is it? Write the name.	Describe it! Words like soft, rough, etc
A colorful object			
A small object			
A soft object			
A shiny object			

Activity 3 Sketch & Spy

Instructions

1. Draw (✏) each object asked in the box and label its color, shape, and texture.
2. Write (✍) one thing you noticed about each object.

a leaf

____

____

a rock

____

____

a piece of fruit

____

____

a feather

____

____

Spot the Observables! (4) Activity

Instructions

1. Below is a list of things you can observe with our senses, and some that require special tools.
2. Identify which ones can be observed by our senses and the ones which require special tools by placing a checkmark (✓) on the correct column.

	Observed with our senses	Observed with special tools
1  smell of a flower	☐	☐
2  movement of the wind	☐	☐
3  size of a tiny germ	☐	☐
4  temperature of soup ***(Adult help is needed!)***	☐	☐
5 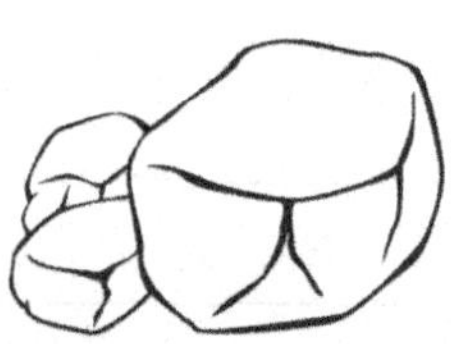 weight of a rock	☐	☐

Activity 5 Can You See It?

Instructions

Answer the following questions by checking (✓) the box of the correct answer.

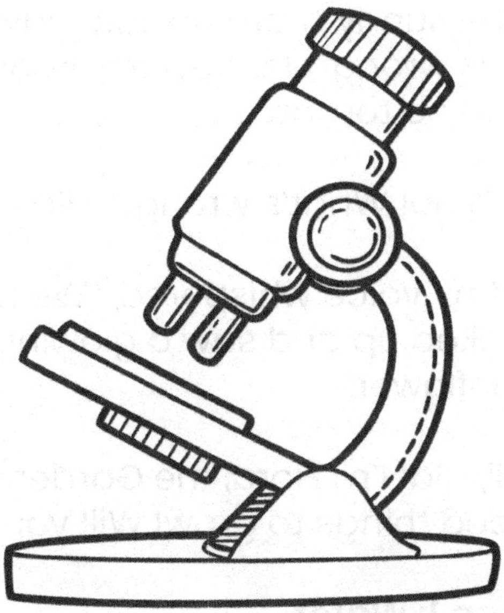

1. **Can you see germs without a microscope?**

   ☐ Yes

   ☐ No

2. **Can you hear the sound of the wind?**

   ☐ Yes

   ☐ No

3. **Can you tell the temperature of an ice cube just by looking at it?**

   ☐ Yes

   ☐ No

4. **Do you need a special tool to measure the length of a pencil?**

   ☐ Yes

   ☐ No

5. **Can you see the stars during the day?**

   ☐ Yes

   ☐ No

Unit 2 Plants and Their Needs

Lila's Magical Garden

6 Activity

One sunny morning, Lila played in her backyard with Milo, her cat. But something was wrong! Her flowers drooped, her grass turned brown, and her tomato plant had no tomatoes.

"Oh no! What's wrong?" Lila asked.

A tiny voice whispered, "We need your help!" Lila looked up and saw a glowing fairy sitting on her sunflower.

"Hi, Lila! I'm Flora, the Garden Fairy. Your plants need things to grow! Will you help me?"

Step 1: Water
Flora pointed to the flowers. "Plants need water!" Lila grabbed her watering can and poured water on the dry soil. The flowers perked up!

Step 2: Sunlight
Flora flew to the tomato plant. "Tomatoes need sunlight!" Lila moved the pot to a sunny spot. "Now you can soak up the sun!"

Step 3: Soil
Flora pointed to the brown grass. "Healthy soil helps plants grow." Lila added rich, dark soil. "This will help you get strong!"

Step 4: Air
Flora twirled in the breeze. "Plants need fresh air, just like us!" Lila took a deep breath. "Our garden has plenty of fresh air!"

Soon, the garden looked bright and happy again! The flowers bloomed, the grass turned green, and tiny tomatoes began to grow.

"Great job, Lila!" Flora cheered. "Water, sunlight, soil, and air! Now you know the secret to growing plants!"

Activity 7 Grow a Mystery Plant!

Be a Plant Detective and Discover How It Grows!
Your job is to plant a mystery seed and track its changes each day.
Will it grow tall? Will it have flowers? Let's find out!

Oops, wait! Adult help is needed!

Materials Needed

- A small cup or pot
- Soil
- A mystery seed! (Any easy-to-grow seed)
- Water
- A sunny spot
- Pencil or crayons

STEP 1: Plant Your Mystery Seed!

1. Fill your cup with soil, leaving space at the top.
2. Make a tiny hole (about an inch deep).
3. Drop in your mystery seed!
4. Cover it with soil and water it gently until damp (not too wet!).
5. Place your cup in a sunny spot.

Step 2: Daily Care & Growth Tracking

1. Water your seed daily to keep the soil moist.
2. Check for changes—does it have a tiny sprout? More leaves?
3. Use the Plant Detective Journal to draw or write about your plant's growth during day 1, day 3, day 5 and day 7.

Instructions

Draw (✏)what you see! Does your seed have a sprout? Leaves? A flower?

Day 1

Day 3

Day 5

Day 7

Grow, Plant, Grow!

8 Activity

Oh no! A tiny seed is trying to grow, but it needs sunlight, water, and healthy soil to become a big, strong plant. Can you help it find its way? Be careful—some paths have obstacles like big rocks and hungry bunnies that might slow you down!

Instructions

1. Start at the tiny seed at the bottom of the maze.
2. Find a path to reach all three needs for a plant to grow:
   - **Water** – Helps the plant drink!
   - **Sunlight** – Gives the plant energy!
   - **Healthy Soil** – Helps the plant stay strong!
3. Avoid the obstacles (a hungry bunny, a big rock, or a dry patch of land).
4. Color (✎) the plant once it reaches all three things it needs to grow!

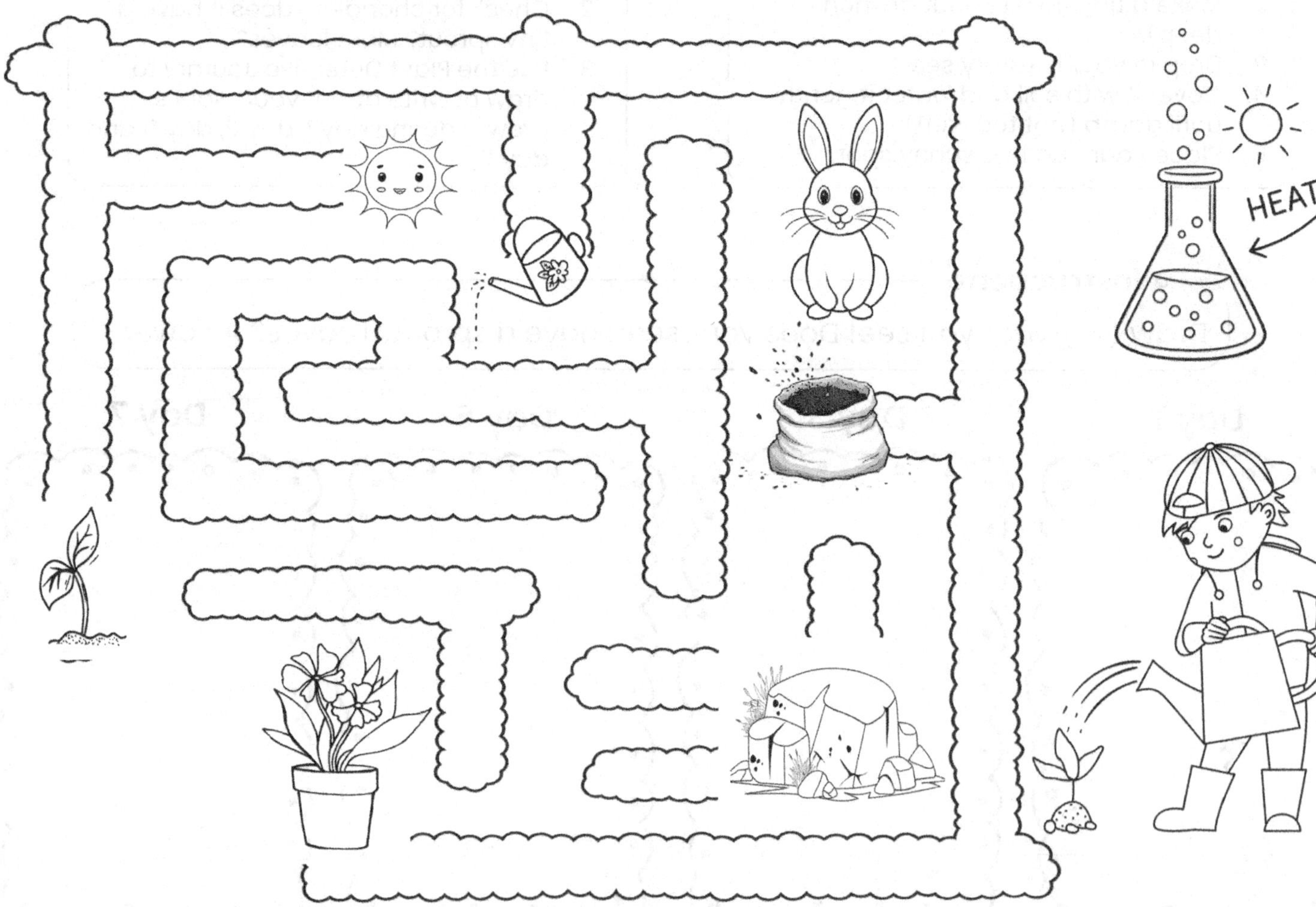

Activity 9 Plant Super Quiz

Plants are amazing! They grow in different places, come in different shapes, and need special things to stay strong.

Instructions

1. Read each plant fact carefully. Identify if the statement is true or false.
2. Color (✎) the leaf of the correct answer.

1 Plants can grow in the dark.

true false

2 Plants need air to live.

true false

3 Plants only need water and nothing else.

true false

4 Soil helps plants grow strong roots.

true false

5 Plants use sunlight to make their own food.

true false

Build-A-Plant Challenge! (10) Activity

Design Your Own Super Plant!
Plants come in all shapes and sizes! Your task is to draw a plant with special powers and label its parts. What will your plant do? Will it grow super tall? Smell amazing? Glow in the dark? Let's find out!

Instructions

1. Draw (✏) a plant in the pot below. It can be a real plant or a crazy, new plant from your imagination!
2. Label the parts of your plant: Roots, Stem, Leaves, Flowers.
3. Write (✍) a fun fact about your plant!
   (Example: "My plant's leaves are huge so it can catch more sunlight!")

Fun fact about my plant!

Life Cycle of Living Things

Activity 11 The Great Transformation

In a quiet pond, a tiny tadpole named Tad wiggled his little tail. He loved swimming in the cool water with his fish friends. But one day, he noticed something strange, his back started to itch!

"Oh no! Am I turning into a fish?" Tad cried.

"No, silly!" laughed Wise Old Turtle. "You are growing legs! That means you're changing into a frog."

Tad wiggled in excitement. Each day, his legs grew longer, and his tail grew shorter. Soon, he could jump on rocks and breathe air. Tad had become a frog!

Not far from the pond, a little caterpillar named Cali munched on a leaf. She loved to eat and eat. One day, she felt very sleepy.

"Why am I so tired?" Cali yawned.

"You are ready for your big change!" said a friendly butterfly.

Cali spun a chrysalis around herself and waited. She stayed inside for a long time. When she finally woke up, she stretched her new wings! She had turned into a beautiful butterfly!

LESSON

Just like Tad and Cali, all living things go through changes as they grow! Frogs start as eggs, then tadpoles, then young frogs, and finally, adult frogs. Butterflies start as eggs, then caterpillars, then form a chrysalis, and finally, become butterflies. Change is a part of life!

Watch Me Grow!

12 Activity

Instructions

1. Color (✏) each stage of the butterfly's life cycle.
2. Draw (✏) arrows to show the correct order of the life cycle.
3. Number each stage from 1 to 4 in the correct sequence.

STAGES TO INCLUDE:

- Egg – A butterfly starts as an egg on a leaf.
- Caterpillar – The egg hatches into a hungry caterpillar.
- Chrysalis – The caterpillar wraps itself into a chrysalis.
- Butterfly – A beautiful butterfly emerges!

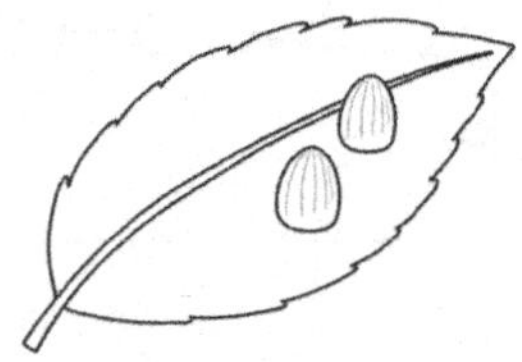

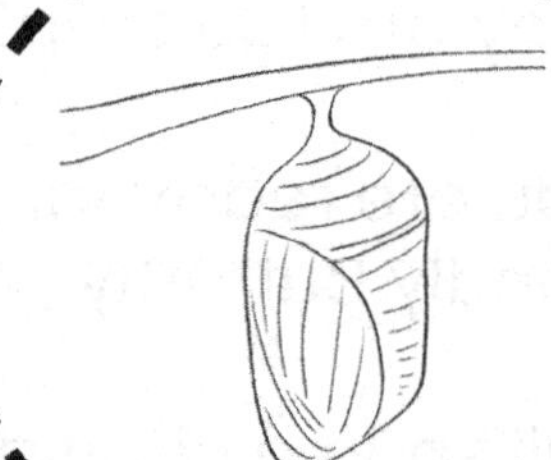

Activity 13 Tad's Big Adventure

Instructions

1. Draw (✏) and color (🖍) each stage of a frog's life in the comic strip panels.
2. Add speech bubbles or captions to describe what's happening at each stage.

STAGES TO INCLUDE:

- Eggs in Water – A frog begins life as tiny eggs floating in a pond.
- Tadpole – The eggs hatch into tadpoles with tails!
- Young Frog – The tadpole grows legs and its tail shrinks.
- Adult Frog – The frog is fully grown and can hop on land!

Life Cycle Check-Up!

14 Activity

Instructions

Complete the sentences below by filling in the missing life cycle words.

Word Bank		
egg	butterfly	young frog
tadpole	caterpillar	chrysalis

1. A butterfly starts as an ________, then turns into a ________ before becoming a chrysalis and then a ________.

2. A frog's life begins as ________, then it hatches into a ________. As it grows, it loses its tail and becomes a ________.

Activity 15 Frog vs. Butterfly

All living things grow, but not all life cycles are the same! Frogs and butterflies both change, but in different ways. Let's compare their journeys from baby to adult!

Instructions

1. Look (👀)at the life cycle words in the word bank.
2. Decide if each belongs to a frog's life, a butterfly's life, or both.
3. Write (✍) them in the Venn diagram where they belong.

Word Bank

tadpole	egg	forms a chrysalis	flies in the air	lives
caterpillar	young frog	hops on land	grows & changes	in water

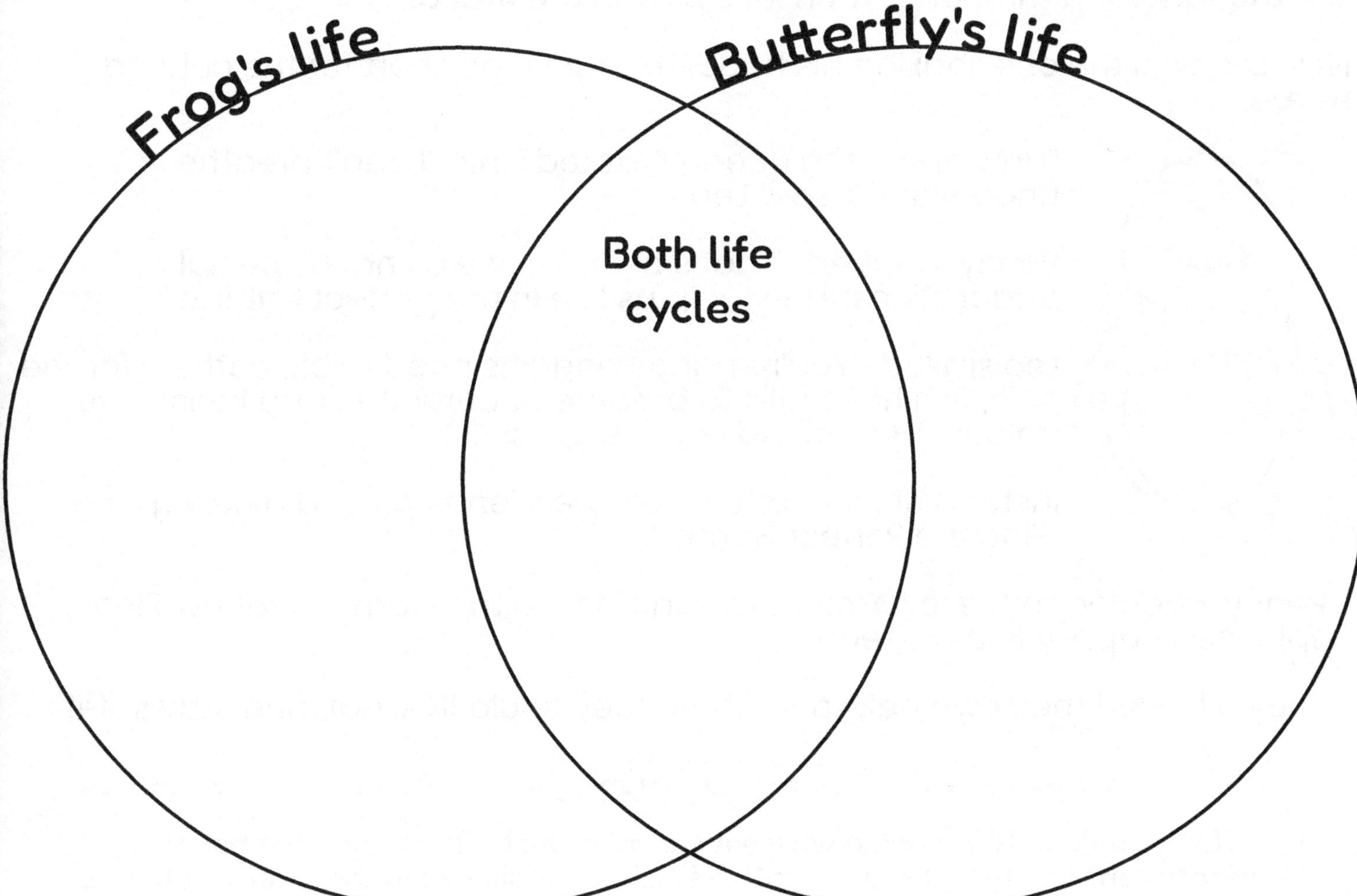

Unit 4 Habitats and Adaptations

The Great Hide-and-Seek! 16 Activity

In the big, wide world, animals live in many different places. Some like it hot, some like it cold, and some love to swim all day!

One day, Leo the Lion wanted to play hide-and-seek with his animal friends.

"Let's play in my home. The hot, grassy savanna!" Leo roared.

"I can't hide there!" said Penny the Penguin. "It's too hot, and I need ice and water to stay cool."

"I can't play there either," said Finn the Fish. "I need water to breathe!"

Leo thought for a moment. "Then let's play in the Arctic!"

Now Leo was the one shaking his head. "My fur is too short for the cold. I'd freeze!"

"How about the ocean?" asked Finn. "I can't breathe underwater!" said Leo.

Penny laughed. "That's because we all have special adaptations! They help us live in the perfect habitat."

Leo smiled. "You're right! Penguins need thick feathers for the cold, fish have gills to breathe underwater, and lions have golden fur to blend into the grass."

Instead of hide-and-seek, the friends played a new game: "Find the Perfect Home!"

Penny waddled to the icy Arctic. Leo ran back to the warm savanna. Finn splashed happily in the ocean.

They all loved their own habitats, where they could live, eat, and stay safe!

LESSON

Every animal has a home where it survives best. This is called a habitat! Adaptations like fur, gills, and feathers help animals live in their perfect home.

Activity 17 Build-A-Habitat!

Animals live in different places. Can you create a habitat for an animal?

Instructions

1. Choose a habitat: Forest, Desert, Arctic, Ocean.
2. Draw (✏) your habitat in the big space below!
3. Add animals that live there (Example: a camel in the desert or a polar bear in the Arctic).
4. Label (✍) at least three things animals need in this habitat: food, water, and shelter.

Adaptation Detective!

18 Activity

Animals have special adaptations that help them survive.
Let's match each animal to its special power!

Instructions

Draw a line to match each animal to its adaptation.

stores fat in its hump for energy

has thick fur to stay warm

can see at night to catch food

has gills to breathe underwater

Activity 19 Who Lives Where?

Instructions

1. Let's test what you know about where animals live!
2. Color the right answer for each question!

1 Which animal lives in the ocean?

dolphin | tiger | elephant

2 Where does an owl make its home?

underground | in a tree | in the ocean

3 Which habitat is best for a snake?

Snowy Arctic | rainforest | ocean

4 Where would you find a kangaroo?

Arctic | Australian Outback | desert

5 Which animal lives in a wetland?

scorpion | frog | camel

Who Belongs Where? 20 Activity

Some animals are lost!
Help them find their correct habitat.

Instructions

1. Look at the animals below.
2. Draw (✏) a line to match each animal to its correct habitat.

camel

desert

clownfish

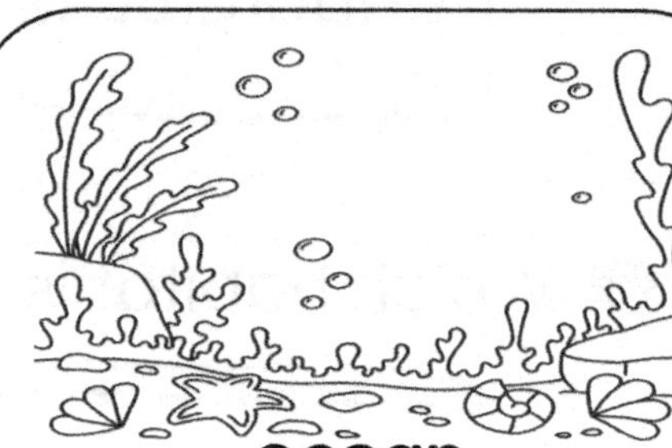

ocean

polar bear

Arctic

monkey

rainforest

Unit 5 Properties of Materials

Activity 21 The Great Material Mystery

One sunny morning, Max the Inventor and his best friend Luna the Scientist were on a mission! They wanted to build the perfect treehouse in Max's backyard. But there was a problem.

"We need the right materials!" Luna said. "Some things are too soft, some are too hard, and some just won't work!"

Max scratched his head. "How do we know what's best?"

Just then, Max's little brother, Ben, ran by with an umbrella. It wasn't even raining!

"Ben! Why are you carrying that?" Max laughed.

"Mom says I can't take my paper hat outside because it'll get wet! But my umbrella is waterproof!" Ben said proudly.

Luna's eyes lit up. "That's it! We need to pick materials based on their properties!"

They started testing materials for their treehouse parts:
For the walls – Wood! It's hard and strong so it won't break.
For the roof – Plastic sheets! They are waterproof and will keep the rain out.
For the seats – Cushions! They are soft and comfy to sit on.
For decorations – Colorful fabric! It's flexible and easy to hang up.

Max grinned. "Now our treehouse will be strong, dry, comfy, AND cool-looking!"

Luna nodded. "See? Knowing the properties of materials helps us make smart choices!"

As they climbed up into their new treehouse, Ben giggled and asked, "Can we make a slide out of paper?"

Max and Luna burst out laughing. "Not unless you want a BIG crash! Paper is too weak!"

And from that day on, Max, Luna, and Ben became Material Detectives, always checking the properties of things around them!

The Great Material Hunt 22 Activity

Instructions

1. Look around your classroom, home, or outside for objects with different properties.
2. Draw (✏) and write (✍) the name of each object you find in the table below.

Property	Write the name of the object found	Draw the object you've found	Where did you find it?
hard			
soft			
waterproof			
flexible			

Activity 23 Sink or Float?

Instructions

1. Gather objects like a coin, a sponge, a leaf, or other small, safe objects you can find with an adult's permission.
2. Predict by putting a checkmark (✓) if each object will sink or float.
3. Drop them into a bowl of water and record what happens!

OBJECT	PREDICTION	WHERE DID YOU FIND IT?
	☐ sink ☐ float	
	☐ sink ☐ float	
	☐ sink ☐ float	
	☐ sink ☐ float	

Melt It or Keep It?

24 Activity

Instructions

1. Find objects that might melt in the sun (like an ice cube, chocolate, or a crayon).
2. Predict what will happen when placed in the sun for 10 minutes.
3. Check and record the results!

Oops, wait! Adult help is needed on this activity!

OBJECT	PREDICTION	WHERE DID YOU FIND IT?
	☐ melt ☐ stay the same	
	☐ melt ☐ stay the same	
	☐ melt ☐ stay the same	
	☐ melt ☐ stay the same	

Activity 25 Material Quiz

Instructions

1. Let's test what you know about the properties of materials!
2. Color (✎) the right answer for each question!

1. Glass is:

soft | hard | flexible

2. Rubber bands are:

hard | flexible | waterproof

3. What material is used to make raincoats?

paper | plastic | metal

4. A sponge is:

soft | hard | rigid

5. Wood is:

flexible | hard | waterproof

States of Matter

The Magic Tea Party

26 Activity

One afternoon, Lily and Ben were playing in the backyard when their Grandma called, "Who wants to help me make tea?" "Me!" they shouted, running inside.

Grandma smiled and said, "Making tea is a perfect way to see the three states of matter in action!"

Solid: The Sugar
Grandma placed a sugar cube on the table. "This is a solid," she said. "It has a shape that doesn't change unless we break it." Ben poked the sugar. "Yup! It stays the same!"

Liquid: The Water
Grandma poured water into the kettle. "This is a liquid," she said. "It takes the shape of whatever container it's in." Lily tilted her head. "That's why my juice fills my cup but doesn't stay the same shape!"

Gas: The Steam
Soon, the kettle started whistling. Steam rose from the spout. "Look! The water is turning into gas!" Grandma said. "It's changing from a liquid to a gas because of the heat!"

Ben waved his hand through the steam. "Whoa! It disappears!"

Grandma nodded. "That's because gases spread out into the air."

After the tea was ready, they stirred in the sugar cube. It disappeared!

"Wait! Where did the sugar go?" Lily asked.

"It dissolved," Grandma explained. "It mixed with the liquid, but it's still there. You can taste it!"

Ben took a sip. "Mmm... science is delicious!"

Everyone laughed as they enjoyed their magical tea party, where solids, liquids, and gases came to life!

Activity 27 State of Matter Sorting

Solids have a fixed shape, liquids take the shape of their container, and gases spread to fill any space!

Instructions

1. Cut out pictures from magazines or newspapers of solids, liquids, and gases.
2. Sort and glue them under the correct column in the table below.
3. Label each picture!

SOLID	LIQUID	GAS

States of Matter Hunt! 28 Activity

Instructions

1. Look around your classroom, home, or outside for examples of each state of matter.
2. Draw (✏) and write (✍) the name of each object you find in the table below.

State of Matter	Write the name of the object found	Draw the object you've found	Where did you find it?
 Solid			
 Liquid			
 Gas			

Activity 29 Shape Shifters

Instructions

1. Look at the list of objects. Think about what happens when you heat them or cool them.
2. Draw (✏) or write (✍) what happens to each one in the table below!

Object	What happens?
 ice cubes	heated →
 chocolate	cooled →
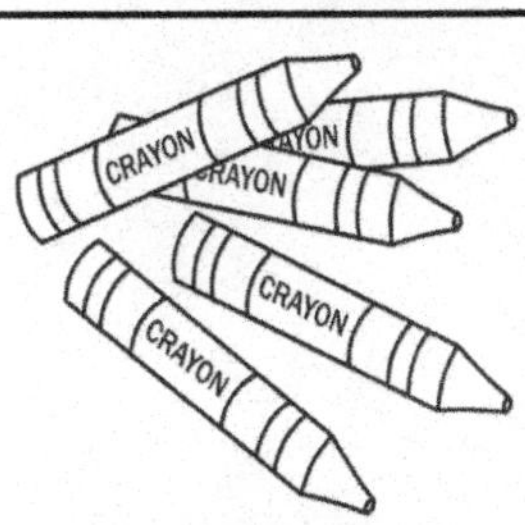 crayons	heated →
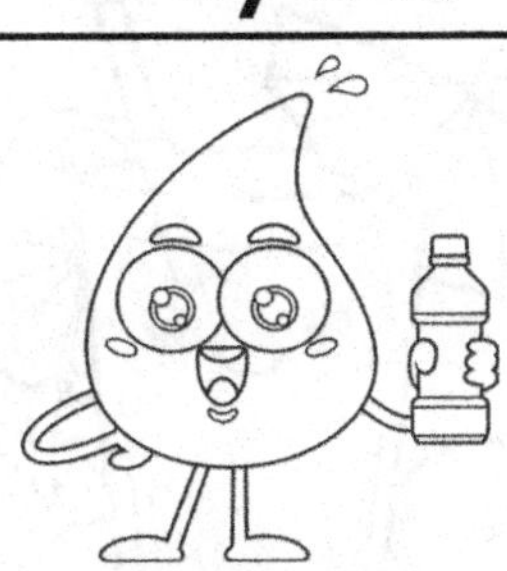 water	cooled →

Matter Quiz

30 Activity

Instructions

Put a checkmark (✓) on the correct answer.

1 Water is solid at room temperature.

- ☐ True
- ☐ False

2 Which of these is a solid?

- ☐ juice
- ☐ rock
- ☐ air

3 Which state of matter fills up any space it is in?

- ☐ solid
- ☐ liquid
- ☐ gas

4 What happens when you put water in the freezer?

- ☐ It turns into gas.
- ☐ It stays the same.
- ☐ It freezes into ice.

5 Steam is a gas.

- ☐ True
- ☐ False

Unit 7

The Water Cycle

Activity 31 Adventures of Drippy the Raindrop!

Drippy the Raindrop was not like any other drop of water. He loved adventures!

One sunny day, Drippy was resting in the ocean, playing with fish and splashing with his water friends. Suddenly, he felt warm and light. "Oh no! I'm getting lighter!" Drippy giggled.

The sun's warm rays made Drippy rise into the sky. "Wheee!" he laughed as he floated up, up, up. This was called **evaporation**, when water turns into gas and floats into the air!

Up in the sky, Drippy met lots of other water drops. They were hugging together to form a big, fluffy cloud. "Hello friends! Now we're in **condensation**!" Drippy cheered.

The cloud got bigger and heavier as more drops joined in. Soon, the cloud turned dark and heavy. "Uh, oh!" Drippy said. "I think it's time to fall back down!"All the water drops tumbled from the cloud as rain. This was called **precipitation**.

Drippy fell onto the top of a tall mountain. "That was a long fall!" he laughed. He rolled down the mountain, joining a river that flowed all the way back to the ocean. "I'm back where I started!" Drippy cheered. This gathering of water is called **collection**!

But guess what? Drippy's adventure wasn't over. It would happen again and again!

"That's the water cycle," Drippy said proudly. "Water is always moving!"

Drippy's Water Cycle Comic 32 Activity

Instructions

1. Draw what happens to Drippy in each stage of the water cycle.
2. Write a sentence under each picture to explain what's happening.

Evaporation

Condensation

Precipitation

Collection

Activity 33 Water Cycle!

Instructions

Divide the circle into four equal parts and draw ( ) and color ( ) each stage of the water cycle.

STAGES OF WATER CYCLE:

- **Evaporation** (Water rises from lakes, rivers, or oceans.)
- **Condensation** (Water vapor turns into clouds.)
- **Precipitation** (Rain, snow, or hail falls from the sky.)
- **Collection** (Water gathers in lakes, rivers, or underground.)

Color-by-Stage Water Cycle 34 Activity

Instructions

1. Look (👀) at the water cycle picture below.
2. Use the color guide below to color (✏) each part correctly.
3. Label (✍) each stage of the water cycle.

Color Guide

- Blue for rivers, lakes, and oceans (Collection).
- Yellow for the sun (Evaporation).
- White for clouds (Condensation).
- Gray for rain and snow (Precipitation).

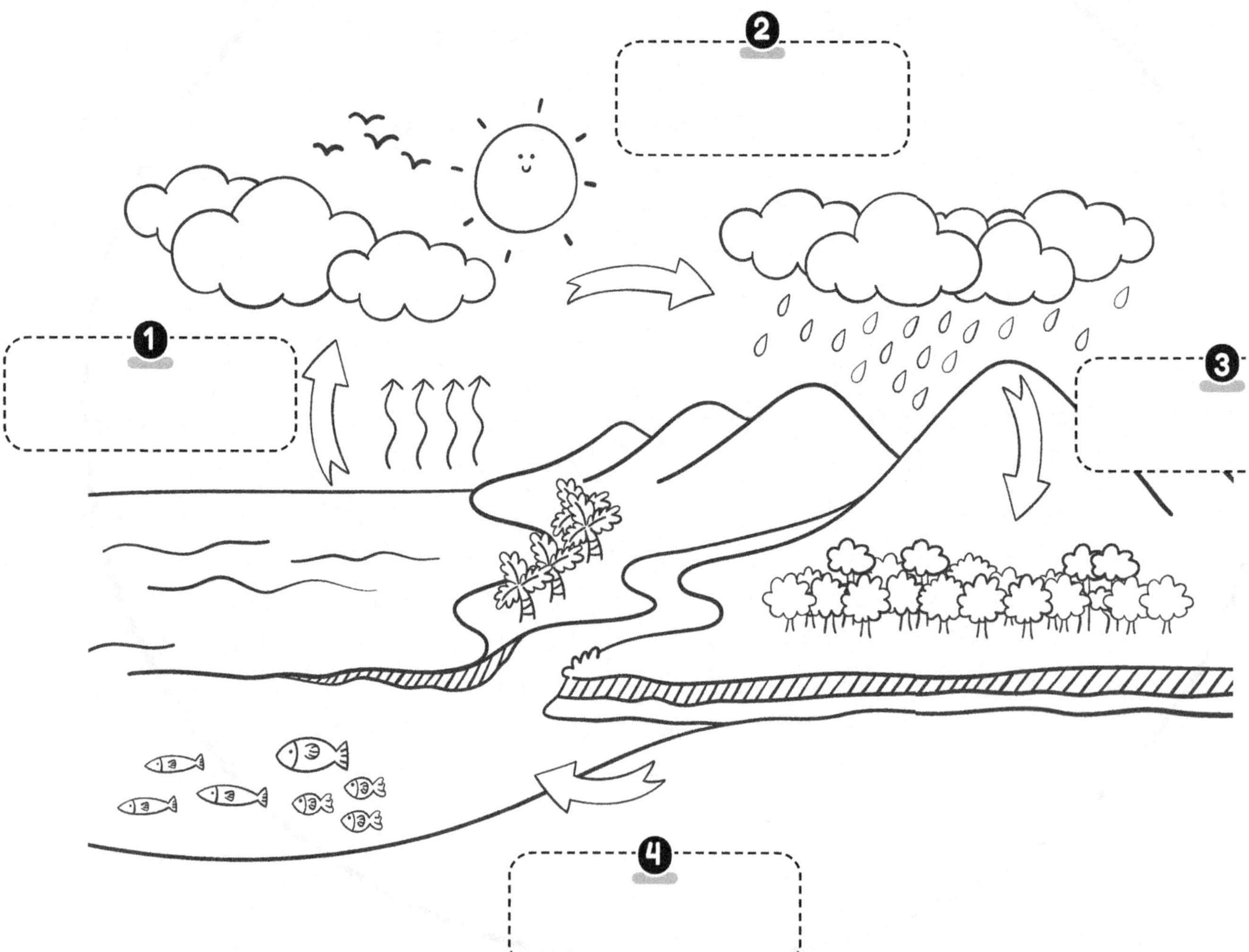

Activity 35 Water Cycle Quiz

Instructions

With the use of your crayon, color (✎) the water drops of the correct answer.

1. Which stage of the water cycle happens when the sun heats up water?
   - Condensation
   - Evaporation
   - Precipitation

2. Clouds are formed during which stage?
   - Evaporation
   - Condensation
   - Collection

3. What happens when clouds get too heavy?
   - They disappear
   - They turn into the sun
   - They rain, snow, or hail

4. Which part of the water cycle refills rivers and lakes?
   - Precipitation
   - Evaporation
   - Condensation

5. Which stage of the water cycle turns water into gas?
   - Collection
   - Condensation
   - Evaporation

Unit 8

Weather and Seasons

The Four Seasons Parade!

36 Activity

Lily and Max loved playing outside, but one day, Lily asked, “Why does the weather keep changing?”

Just then, a magical gust of wind whooshed around them! A friendly voice called out, “Welcome to the Four Seasons Parade!”

First, came **Spring**, dancing in with blooming flowers and chirping birds. “I bring rain and sunshine so plants can grow!”

Then, came **Summer**, waving with bright, golden rays. “I bring hot, sunny days for swimming and ice cream!”

Fall twirled in next, covered in red and orange leaves. “I bring cool winds, falling leaves, and back-to-school fun!”

Last came **Winter**, wrapped in a cozy scarf. “I bring chilly air, snowflakes, and warm hot chocolate!”

The wind swirled again, and the seasons lined up, saying, “Together, we make the whole year special!”

Max grinned. “Wow! So that’s why the weather keeps changing!”

Lily laughed. “Let’s enjoy every season’s parade!”

Activity 37 Color the Seasons Parade!

Instructions

1. Color (✎) each season using the suggested colors below.
2. Write (✍) one fun thing you do in each season.

SUGGESTED COLORS:

- Spring – Green and pink (flowers and new leaves)
- Summer – Yellow and blue (sunshine and beach fun)
- Fall – Orange and red (falling leaves)
- Winter – White and blue (snowy days and warm clothes)

Spring

____

____

____

Summer

____

____

____

Fall

____

____

____

Winter

____

____

____

My Weather Log!

38 Activity

Instructions

1. Observe the weather outside every day for six days.
2. Draw (✏) a symbol (see the symbol guide provided) that shows the weather for each day.
3. Write (✍) about what you see.

SYMBOL GUIDE:

Day 1

Day 2

Day 3

Day 4

Day 5

Day 6

Activity 39 Season Scene Drawing!

Instructions

1. Choose one season (spring, summer, fall, or winter).
2. Draw (✏) a picture of what the season looks like! Include:
   * Trees & plants (Are they bare, green, or colorful?)
   * Clothes (Are people wearing coats or shorts?)
   * Activities (Are kids swimming or building a snowman?)
3. Write (✍) a sentence describing your season!

Weather & Seasons Quiz! 40 Activity

Instructions

With the use of your crayon, color (✎) the correct answer.

1. Which season is the hottest?

winter | spring | summer

2. What do we use on rainy days?

umbrella | sunglasses | mittens

3. Which season do leaves change color?

summer | fall | spring

4. Which season is it most likely to snow?

summer | winter | spring

5. What does a thermometer measure?

temperature | rainfall | wind speed

Unit 9 Simple Machines

Activity 41 The Super Tool Team's Rescue!

Every night in Inventorville, the Super Tool Team came to life! Each tool had a special superpower to make work easier.

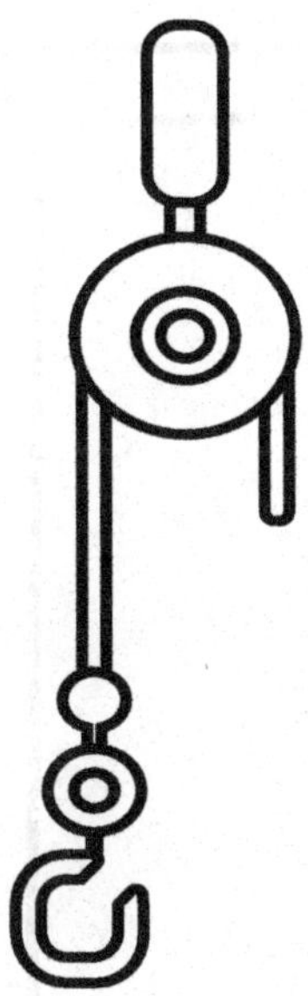

Lever Louie lifted heavy things like a seesaw or crowbar.

Wheelie Wanda helped things roll like a bike or shopping cart.

Pulley Pete lifted heavy loads using ropes and wheels.

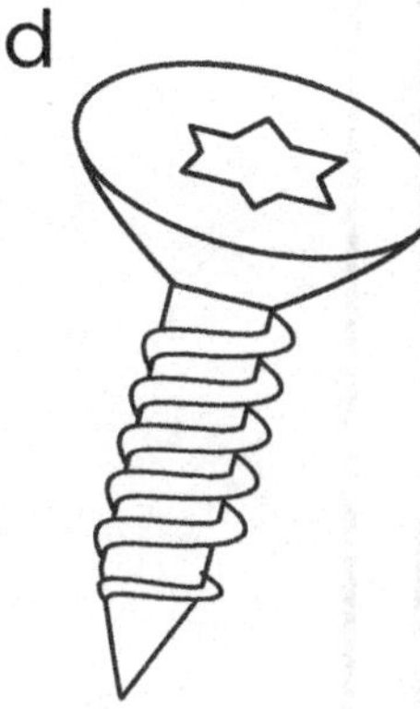

Incline Ivy made it easy to move things up a slope.

Wedge Walter split things apart like a knife or axe.

Screw Sally held things together tight.

One morning, a giant crate blocked Main Street! The mayor shouted, "How can we move it?"

The Super Tool Team worked together: Lever Louie lifted one side. Wheelie Wanda rolled it with wheels. Pulley Pete pulled it up. Incline Ivy made a ramp. Wedge Walter kept it from rolling too fast. Screw Sally tightened it shut.

With a final push, the crate rolled away! "Simple machines make work easier!" cheered the town. The Super Tool Team smiled and went back to their places, ready for the next adventure!

Be an Inventor!

42 Activity

Instructions

1. Draw (✏) a new simple machine that you would invent!
2. Label (✍) its parts.
3. Answer the question below.

What problem does your machine solve?

Activity 43 Find the Simple Machines!

Instructions

Write (✍) the correct type of simple machine that each object corresponds to.

Types of Simple Machines		
pulley	inclined plane	wedge
lever	wheel and axle	screw

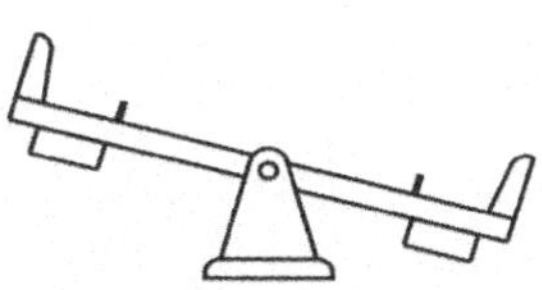

seesaw

____

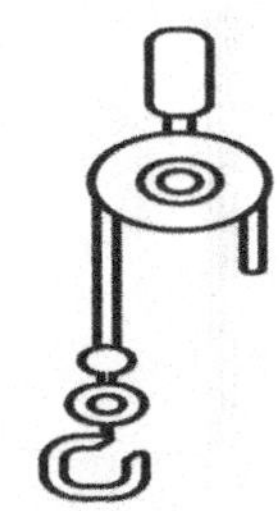

pulley

____

bicycle

____

knife

____

ramp

____

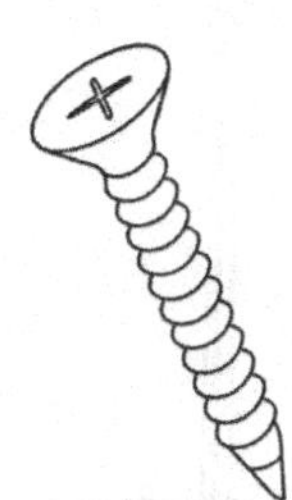

screw

____

Machines in the Real World! 44 Activity

Instructions

1. Walk around your home, school, or playground.
2. Find an example of each simple machine.
3. Draw (✏) or write (✍) what you found!

lever	pulley	inclined plane

wedge	screw	wheel and axle

Activity 45 Test Your Machine Skills

Instructions

With the use of your crayon, color (✎) the correct answer.

1. A lever is used to:

cut things	lift things	roll things

2. Which simple machine helps lift a bucket from a well?

pulley	wedge	screw

3. What simple machine is a doorstop?

wedge	wheel and axle	pulley

4. What is the purpose of an inclined plane?

to make lifting easier with a slope	to spin wheels	to hold things together

5. A flagpole uses which simple machine?

wedge	pulley	lever

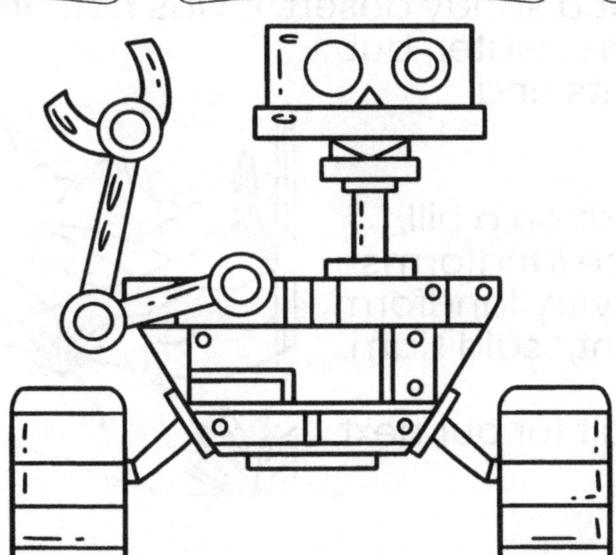

Unit 10

Earth's Landforms

The Great Landform Adventure! 46 Activity

One day, Liam and Lily packed their backpacks for a great adventure to explore different landforms!

First, they climbed a tall mountain where the air was cool and crisp. "Wow! Look how high we are!" Liam said.

Then, they walked down into a deep valley surrounded by tall trees. "It feels cozy down here," said Lily.

Next, they followed a winding river that flowed over rocks and under bridges. "The river helps plants and animals get water," explained Liam.

After a long walk, they reached a wide, flat plain covered with grass and flowers. They ran through the fields, feeling the wind in their hair.

Finally, they arrived at a sandy desert. It was hot, and the ground was dry. "This place has less water, but it's full of special plants and animals," Lily said.

As the sun set, they sat on a hill, looking out over all the landforms they had explored. "Every landform is special and different," said Liam.

Lily smiled. "I can't wait for our next adventure!"

Activity 47 Match the Landforms!

Instructions

Draw (✏) a line from the picture to the correct landform name, and a separate line from the name to its description.

	mountain	a dry place with lots of sand
	hill	a tall, rocky landform that rises high
	desert	a raised area of land that is smaller than a mountain
	river	a flat area with grass and plants
	plain	a flowing body of water

Draw Your Own Adventure 48 Activity

Instructions

1. Draw (✎) a landscape that includes a mountain, river, valley, and plain.
2. Use different colors (✎) for each landform.
3. Label (✍) each landform and add small details like trees, rocks, or animals.

Activity 49 Landforms Word Search

Instructions

1. Find the hidden words in the letter grid.
2. Words may be in diagonal, horizontal or vertical.
3. Circle (○) or highlight (✎) each word as you find it!

D	H	U	V	Y	L	H
E	R	P	A	K	A	I
S	I	V	L	D	K	L
E	V	B	L	A	E	L
R	E	B	E	S	I	R
T	R	F	Y	K	M	N
O	C	E	A	N	O	H

WORD LIST:

- **RIVER** – A flowing body of water.
- **OCEAN** – A vast body of saltwater.
- **PLAIN** – A wide, flat land.
- **VALLEY** – A low area between mountains or hills.
- **DESERT** – A dry, sandy land with little water.

Test Your Landform Knowledge 50 Activity

Instructions

With the use of your crayon, color (✎) the correct answer.

1. Which landform is the tallest?

valley	mountain	plain

2. Which landform has deep, flowing water?

river	desert	hill

3. What landform is covered in sand and gets very little rain?

forest	ocean	desert

4. Which landform is flat and has a lot of grass?

plain	mountain	beach

5. A valley is usually found between________________

mountains	oceans	deserts

Unit 11 Classifying Animals

Activity 51 The Mystery Zoo Escape!

One morning, Zookeeper Mia woke up to a big problem! The animals had escaped from their habitats!

She grabbed her notebook and rushed outside. “Oh no! The gates are open, and the animals are in the wrong places!”

First, Mia ran to the desert exhibit. She saw a monkey swinging from a cactus!
“Monkeys belong in the rainforest, not the desert!” Mia laughed. The monkey agreed and climbed onto her shoulder.

Next, Mia heard splashing in the rainforest. There was a penguin waddling in the trees! “Penguins need cold places!” Mia said. The penguin shivered and climbed into her cart.

Mia then went to the pond and saw a lion splashing in the water!
“Lions belong in the savanna, not the pond!” Mia giggled. The lion shook its mane and climbed into the cart.

Finally, Mia checked the Arctic habitat. To her surprise, an alligator was lying on the snow! “Alligators need warm, swampy places, not ice!” Mia explained. The alligator shivered and joined the others in the cart.

Mia took all the animals back to their homes:
The monkey swung in the rainforest.
The penguin slid into icy water.
The lion stretched under the savanna sun.
The alligator splashed in its swamp.

“Phew! What a day!” Mia smiled.

Just then, she heard a noise behind her. A camel was standing in the rainforest! “Oh no, not again!” Mia laughed. Time for another adventure!

Animals to Their Homes

52 Activity

Did you know that animals are grouped by their body features, how they move, and where they live?
Some animals have fur, some have scales, and some even have feathers!
Some live on land, others live in water, and some live in both!

Instructions

Draw (✎) a line to match each animal to its correct habitat.

 monkey	savanna
 penguin	rainforest
 lion	Arctic
 alligator	desert
 camel	swamp

Activity 53 Animal Bingo

Instructions

1. Read each clue and mark off the correct animal trait.
2. When you hear a trait that matches your bingo card, color (✐) that square!
3. Get 5 in a row (up, down, or across)? Yell "BINGO!"

 has fur	 can swim	 has wings	 lives in trees	 no bones
 lays eggs	 cold body	 lives in water	 has 4 legs	 has a shell
 has scales	 mammal	 eats meat	 can run fast	 can fly
 free space	 warm body	 blends in	 lives in desert	 climbs trees
 eats plants	 sharp teeth	 has gills	 can jump	 loves snow

- "I flap my wings and fly in the sky!"
- "I swim in a pond and breathe through gills."
- "I have no bones and a squishy body."
- "I use my big teeth to eat other animals."
- "I'm soft and furry. I like to cuddle!"
- "I'm cold-blooded and I slither on the ground."
- "I have a shell and move really slow!"
- "I live in snowy places like the Arctic!"
- "I swing in trees and love bananas!"

Create Your Own!

54 Activity

Instructions

1. Invent a new animal!
2. Draw (✎) it in its habitat.
3. Write (✍) 3 facts about what it eats, where it lives, and how it moves.

Write (✍) 3 facts about your animal.

Activity 55 Test Your Animal Smarts!

Instructions

With the use of your crayon, color (✎) the correct answer.

1. Which animal lives in the ocean?

lion	dolphin	eagle

2. Which animal has a hard shell?

rabbit	turtle	kangaroo

3. Which animal can fly?

penguin	owl	elephant

4. Which animal is cold-blooded?

snake	dog	monkey

5. Where do camels live?

ocean	desert	jungle

Erosion and Earth Changes

Rocky and the Rolling Rain 56 Activity

Meet Rocky the Rock, who lived proudly on top of a big mountain. He was round, strong, and loved the sunshine.

But one day, dark clouds gathered and the rain came pouring down.

"Whoa! What's happening?" Rocky shouted as the water rushed past him.
Drip, drip, whoosh! The rainwater washed away tiny bits of soil and pebbles around him.

"Hey! My mountain is changing!" Rocky said in surprise.

The next day, the wind came howling in. It blew dust and sand against Rocky, slowly smoothing his edges.

"That tickles!" he laughed. "But I think I'm getting smaller!"

Then came the river, tumbling and splashing down the mountain. It carried pieces of the mountain far, far away.

"Where are you taking my friends?" Rocky called.

A wise old tree nearby whispered, "That's called erosion, Rocky. Water, wind, and ice change the land slowly over time."

Rocky blinked. "Wow! The Earth changes even when I'm just sitting still!"

And from that day on, Rocky kept watching as the land around him changed, little by little, day by day.

KEY IDEA

Erosion is when wind, water, or ice moves rocks and soil and slowly changes the shape of the land.

Activity 57 Earth Gets a Makeover!

Erosion is how wind, water, and ice slowly wear away rocks, sand, and soil. It can change the shape of the land, one tiny piece at a time!

Instructions

1. Draw (✏) a beach, mountain, or hillside in the first box – before erosion.
2. In the second box, draw (✏) what it might look like after a storm, wind, or heavy rain!
3. Answer (✍) the questions below.

before the erosion	after the erosion

What happened to the sand, soil, or rocks?

What caused the change?

Rock Stars: Nature's Treasures! 58 Activity

Rocks come in all shapes, colors, and sizes—and each one tells a story!

Instructions

1. Go outside and collect 3 interesting rocks.
2. Tape or draw (✏) a picture of each one.
3. Write (✍) about them by describing their color, texture, size, and shape.

rock 1

rock 2

rock 3

Activity 59 Watch It Wash Away!

Instructions

1. Fill a tray with sand or dirt.
2. Pour water slowly from a cup and watch what happens.
3. Draw (✏) or describe what the water did.
4. Write (✍) your observation by answering the questions below.

before water	after water

What changed?

Where did the soil go?

What Do You Know About Erosion? 60 Activity

Instructions

Put a checkmark (✓) on the correct answer.

1. What is erosion?
   - ☐ a plant
   - ☐ When land wears away
   - ☐ A kind of rock

2. Which of these can cause erosion?
   - ☐ roots
   - ☐ wind
   - ☐ sunlight

3. Which slows down erosion?
   - ☐ Plants with roots
   - ☐ More wind
   - ☐ Taking away rocks

4. What happens to rocks during erosion?
   - ☐ They grow
   - ☐ They melt
   - ☐ They break down

5. Which place might erode fastest?
   - ☐ A grassy field
   - ☐ A bare hill with no plants
   - ☐ A forest

Unit 13 Exploring Magnetism

Activity 61 Max and the Mysterious Magnet!

One sunny afternoon, Max the curious raccoon was digging through the forest when he found a strange shiny stick with a red tip.

"Hmm... what's this?" Max wondered. He nudged it toward a pile of old buttons, and CLINK! They jumped right onto it!

"Whoa! It's magic!" Max squeaked. "This stick is pulling things without touching them!"

Just then, Ollie the Owl flew down and chuckled, "That's not magic, Max! That's a magnet! Magnets can pull certain metal objects like paperclips, nails, or even steel screws!"

Max blinked. "So it's like a superhero with invisible powers?"

"Exactly!" Ollie hooted. "Magnets have a special force that attracts things made of iron or steel. But not everything sticks, try it on a leaf!"

Max waved the magnet over a leaf. Nothing happened.
"But it worked on the button!" he said.

"Not all materials are magnetic," said Ollie. "Time to explore and see what sticks and what doesn't!"

Max grinned and ran off into the forest, ready to test his magnet on rocks, sticks, cans, and toys.

"Let's find out what this magnet really loves to stick to!"

KEY IDEA

Magnets attract some materials (like iron or steel) and don't work on others (like plastic or wood). They can pull things without even touching them!

What Sticks? Magnet Hunt! 62 Activity

Instructions

1. Search for different items around your home or classroom.
2. Test the items using a magnet.
3. Draw (✎) the items on the box and write the name of each item on the blanks.
4. Circle (○) the thumbs up icon if the item is magnetic and thumbs down if not.

Draw the item.	What is the item?	Is it magnetic?
		yes no
		yes no
		yes no
		yes no

Activity 63 Magnet Maze Mission!

Instructions

1. Place a paperclip at the start of the maze (drawn on paper).
2. Help the rabbit find its way to its food by Using a magnet underneath the page to guide the paperclip to the finish line.
3. Don't touch the paperclip with your hands!

Sort It Out – Magnetic or Not? 64 Activity

Instructions

1. Look at the pictures of different items.
2. Write (✍) "***Magnetic***" or "***Not Magnetic***" next to each.
3. Circle the items you think are made of metal.

paper clip

sock

key

leaf

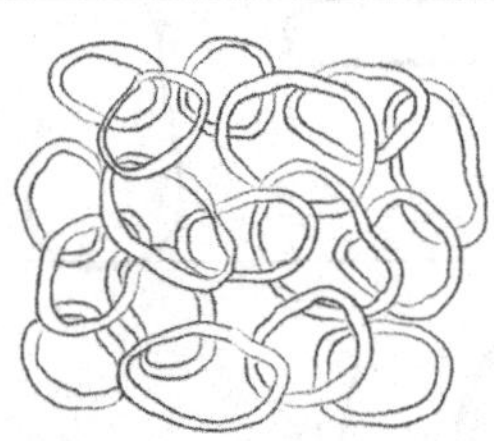

rubber band

tin can

spoon

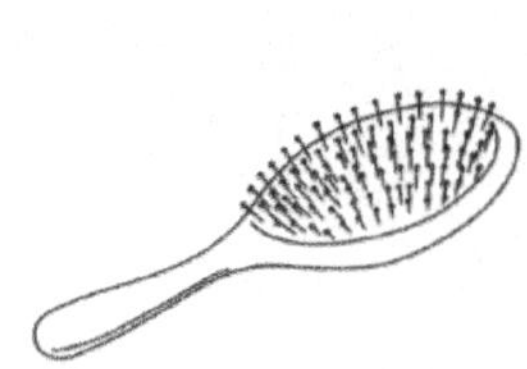

hairbrush

Activity 65 Magnet Mastermind Quiz

Instructions

Circle ***True*** if the statement is true and ***False*** if the statement is false.

1. **Magnets can attract metal objects like paperclips.**
   - true
   - false

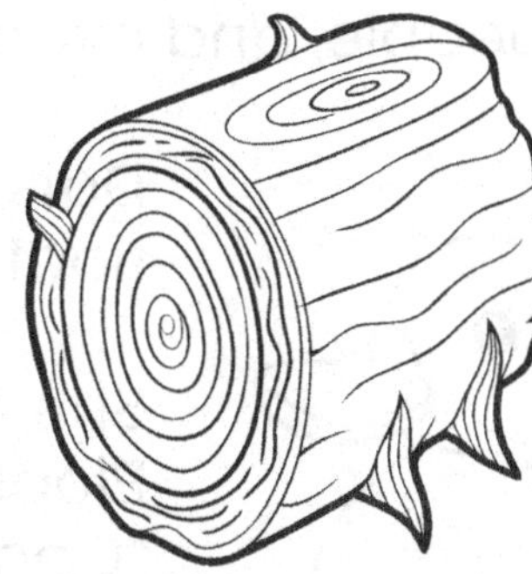

2. **Plastic and wood are magnetic materials.**
   - true
   - false

3. **Magnets only come in one shape.**
   - true
   - false

4. **A compass uses a magnet to help show direction**
   - true
   - false

5. **Magnets only work at night.**
   - true
   - false

Unit 14 Density and Buoyancy

Daisy's Sink or Float Adventure! 66 Activity

One sunny afternoon, Daisy the Duck was waddling near the pond when she spotted a pile of treasures left by the forest animals. An apple, a spoon, a feather, a pebble, and even a toy boat!

Daisy was curious. "Hmm... I wonder which of these will float like me and which will sink to the bottom?"

She gently pushed the apple into the water.
Plop! It floated!
"Cool! The apple stays on top," Daisy quacked.

Next, she dropped the spoon.
KER-PLUNK! It sank straight down.
"Whoa! That one didn't float at all," she said.

Just then, Ollie the Owl flew by. "Daisy, you're testing density and buoyancy!" he hooted.

"Buoyancy? What's that?" asked Daisy.

"It's what makes things float. If something is less dense than water, it floats. If it's heavier for its size, it sinks," explained Ollie.

Daisy smiled and tried the rest. The feather floated, the pebble sank, and the toy boat floated proudly.

"Now I'm a real scientist!" Daisy quacked. "Let's see what else I can test!"

And off she went, waddling with a new mission. The great sink or float experiment!

KEY IDEA

Buoyancy helps things float, and density helps us understand why some things sink and others don't—all depending on how heavy or light they are for their size!

Activity 67 Sink or Float?

Instructions

1. Look at the list of objects.
2. Predict whether each will sink or float in water.
3. Test the objects and record what really happened!

Object	My Prediction *sink or float*	What happened?
 rock		
 leaf		
 feather		
 coin		
 toy boat		
 plastic spoon		
 apple		
 crayon		

Oil vs. Water – A Layered Surprise! 68 Activity

Instructions

1. Pour water into a clear jar.
2. Add oil slowly on top.
3. Watch what happens! Draw (✏) and label (✍) the jar showing which liquid is on top and which is on the bottom.

Oops, wait! Adult help is needed!

Activity 69 Sinker or Floater?

Instructions

1. Draw (✏) 2 objects that sink and 2 objects that float.
2. Label (✍) each one.

sinker object

sinker object

floater object

floater object

Float or Sink? Let's Think! 70 Activity

Instructions

Circle (○) the correct answer.

1. A rock sinks in water because it is
   (a) big
   (b) light
   (c) heavy

2. A feather floats in water because it is
   (a) heavy
   (b) light
   (c) bumpy

3. What is buoyancy?
   (a) The color of water
   (b) The ability to float
   (c) How water tastes

4. What happens if something is more dense than water?
   (a) It floats
   (b) It sinks
   (c) It turns red

5. What does "float" mean?
   (a) It moves to the bottom
   (b) It stays on top
   (c) It disappears

Unit 15

Energy and Light

Activity 71 Luna and the Lost Light

One cloudy evening, Luna the Firefly woke up ready to glow and play with her friends. But something was wrong. Her light wouldn't shine!

"Oh no! Where did my light go?" Luna buzzed. She zipped through the meadow, looking for help.

First, she flew to Sunny the Sunflower.

"Sunny, do you have any light I can borrow?"

Sunny smiled, "I get my light from the sun during the day. That's light energy! But now the sun is sleeping."

Next, Luna flew to Ben the Bunny, who was reading under a glowing lamp. "Ben! How is your lamp so bright?"

"It's plugged into the wall," Ben said. "That's electrical energy. It turns into light energy so I can see my book!"

Luna zipped off again and saw kids holding flashlights.

"Even batteries can give us energy!" Luna cheered. "That's stored energy that turns into light!"

Just then, a little spark started glowing at the tip of her tail.
"My light is back!" Luna laughed. "Energy is everywhere! It just changes from one kind to another!"

And off she flew, lighting up the night sky. One twinkle at a time!

KEY IDEA

Light energy helps us see things and can come from the sun, electricity, batteries, or even fireflies! Energy can change form, like electricity turning into light.

Shadow Watchers!

72 Activity

Instructions

1. Place a small toy or object on a sheet of paper in the sunlight during the day and trace its shadow using a pencil.
2. Come back in the afternoon and trace the shadow again.
3. Write (✍) how the shadow changed and why you think that happened.

morning	afternoon

The shadow changed because...

Activity 73 Power Up the Solar Panel

Instructions

1. Help the sun travel through the maze to reach the solar panel.
2. Avoid objects that block light (clouds, walls) and pass through clear paths (glass, windows).
3. Use a colored pencil (✏) to trace the path of light.

The light couldn't go through the wall because

What Gives Us Light? 74 Activity

Instructions

1. Look at the pictures of objects below.
2. Sort them into "Light Sources" and "Not Light Sources" by coloring (✏) them using the color code given.

Color Code:	
Light Sources - yellow	**Not Light Sources - blue**

rock

sun

chair

cloud

firefly

flower

flashlight

candle

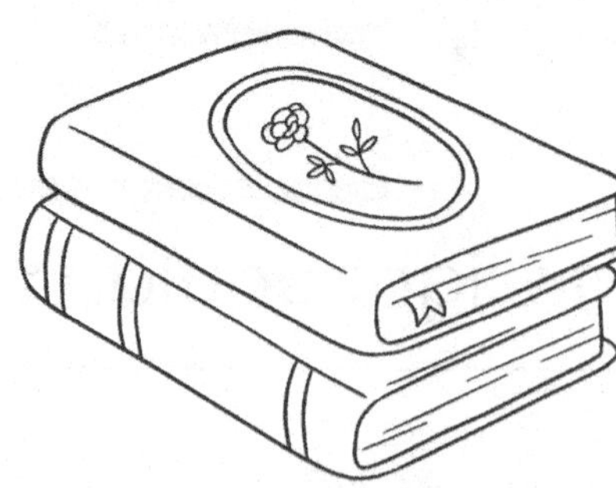
books

Activity 75 What Do You Know About Energy?

Instructions

With the use of your crayon, color (✎) the correct answer.

1. Which is a source of light?

sun | rock | cloud

2. What gives off both light and heat?

ice | fire | fan

3. What helps you see in a dark room?

blanket | lamp | pillow

4. The sun gives off:

music | heat and light | food

5. Which one is a natural source of light?

candle | lightbulb | firefly

Unit 16 Conservation and Natural Resources

Nia the Nature Helper 76 Activity

Nia was a bright and curious girl who loved being outside. She played in the park, splashed in puddles, and listened to the birds sing. One sunny afternoon, she noticed something strange

There was trash near the pond where the ducks swam. The fountain wasn't working because there was no water. A tree had lost too many leaves, and some plants looked tired.

"Oh no," Nia said, "what's happening to nature?"

Just then, Captain Green, a tiny squirrel in a leaf cape, appeared on a branch.

"Hi, Nia! I'm here to teach you about natural resources, like water, trees, air, and soil. We use them every day, but we must protect them."

"Can I help?" Nia asked.

"Of course!" said Captain Green. "When we turn off the faucet while brushing our teeth, we save water. When we plant a tree or reuse paper, we protect forests. When we throw trash in the bin, we keep the Earth clean!"

From that day on, Nia became a Nature Helper. She reminded her family to recycle, helped plant flowers at school, and even made a sign that read: "Take care of nature and it will take care of YOU!"

KEY IDEA

Natural resources are things we get from nature—like water, trees, and air, and we need to use them wisely to keep the Earth clean and healthy!

Activity 77 Think Green!

We use water, air, trees, soil, and energy every day and they're called natural resources. But these resources can run out if we're not careful. That's why conservation is important. It means using only what we need and taking care of the planet!

Instructions

Match each action to the correct impact (what it helps save) by drawing (✏) a line to connect.

ACTION	WHAT IT HELPS SAVE
turning off lights	soil health
recycling paper	energy
composting food scraps	trees
planting trees	fuel
biking instead of driving	cleaner air
picking up trash	water
reusing shopping bags	clean environment
turning off the faucet	plastic

What Comes from Nature? 78 Activity

Instructions

1. Look at the items in the boxes.
2. Color (✎) the ones that come from direct natural resources.

tree

juice

carrot

couch

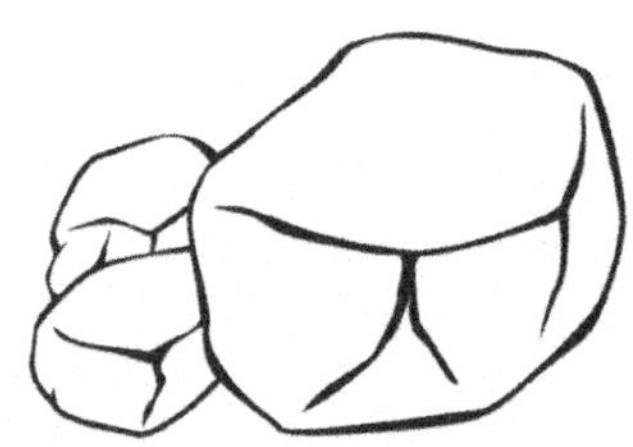

rock

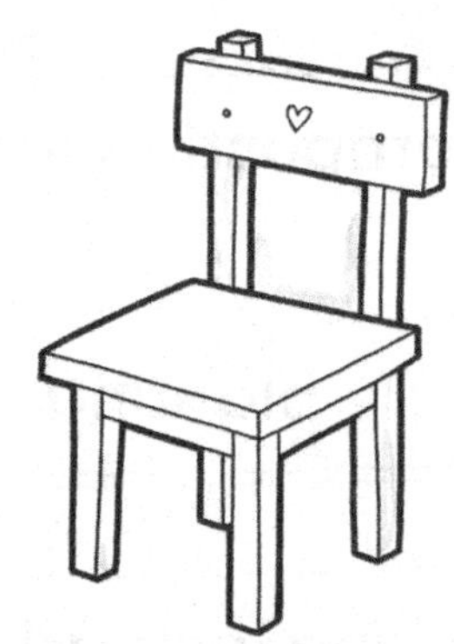

wooden chair

cloud

leaf

battery

Activity 79 My Conservation Pledge

Instructions

1. Write (✍) or draw (✏) 3 things you can do to protect natural resources.
2. Use pictures or words to show your ideas clearly.
3. Sign (✍) your name at the bottom to make it official!

I can save water by...

I can save energy by...

I can help nature by...

"I promise to take care of the Earth by doing these things!"

____

name

True or False? Save Our Earth! 80 Activity

Instructions

Circle ***True*** if the statement is true and ***False*** if the statement is false.

1 **We should throw trash into rivers and oceans.**

true

false

2 **Reusing a lunchbox instead of using paper bags helps reduce waste.**

true

false

3 **The Earth has unlimited water and trees.**

true

false

4 **Turning off electronics when not in use helps save energy.**

true

false

5 **Recycling means throwing things away.**

true

false

Unit 17

Earth's Water Sources

Activity 81 Wally the Water Drop's Adventure

Wally the Water Drop lived in a big, blue ocean. He loved splashing with his fish friends and riding the waves. But one day, Wally got curious.

"Where else can water go?" he wondered.

So, he set off on a big adventure!

First, he floated up into a cloud and became part of the sky!

Then, PLOP! He fell down as rain into a lake where ducks paddled by. From there, he flowed into a river, twisting and turning past trees and rocks.

Next, Wally sank into the ground and found himself in underground water, hiding beneath the soil.

"Wow! Water is everywhere!" Wally said.

He even saw water frozen on top of mountains as snow and felt himself being stored in a big water tank at a school!

Wally smiled. "I may be small, but I'm part of the whole world. People, animals, and plants all need me!"

From oceans to raindrops, Wally learned that water is one of Earth's most important treasures.

KEY IDEA

Earth's water is found in many places. Oceans, rivers, lakes, underground, rain, snow, and even in clouds! We use water to drink, wash, grow food, and keep everything alive.

Color the Blue Planet!

82 Activity

Instructions

1. Turn the book sideways, then look at the world map.
2. Color (✐) all the oceans, lakes, and rivers using ***blue***.
3. Use ***green*** for land to help the water stand out!

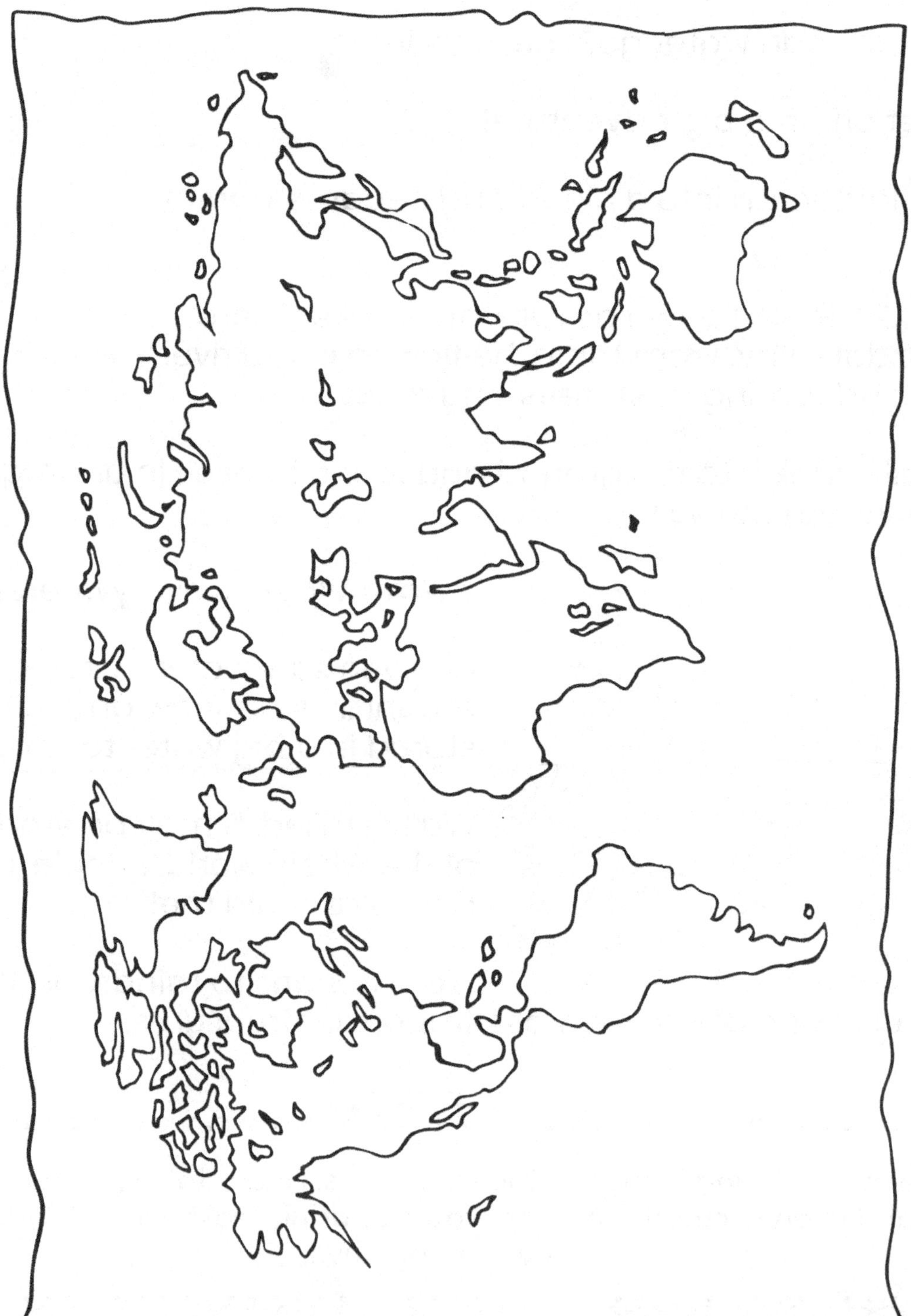

Activity 83

Where's the Water?

Instructions

1. Draw (✏) a line to match each picture to its correct name.
2. Color (🖍) the ones that contain fresh water.

waterfall

pond

river

ocean

Sort That Splash!

84 Activity

Instructions

1. Read the list of water sources and objects.
2. Sort them into the correct categories: Natural or Man-Made water sources.

Water Sources and Objects

ocean	rain	bath tub	sprinkler	swimming pool
faucet	lake	pond	glacier	river

natural water sources	man-made water sources

Activity 85 Water Wise!

Instructions

Circle ***True*** if the statement is true and ***False*** if the statement is false.

1. Oceans are the largest source of water on Earth.
   - true
   - false

2. **Lakes are made of salty water.**
   - true
   - false

3. **We can drink water straight from the ocean.**
   - true
   - false

4. **Some water is stored underground.**
   - true
   - false

5. **Rain is a natural source of water.**
   - true
   - false

Unit 18 Animal Life in Extreme Environments

Arlo the Explorer! 86 Activity

Arlo the Penguin was not your average penguin. He was an animal explorer!

He wore a tiny backpack, a hat, and carried a map. One day, he decided, "I want to meet animals that live in the most extreme places on Earth!"

First, Arlo waddled to the icy Arctic, where he met Polly the Polar Bear. "Brrr!" said Arlo. "How do you live here?" "I have thick fur and fat called blubber to keep me warm!" Polly grinned.

Next, Arlo zipped to the hot, dry desert, where he saw Cammy the Camel. "Whoa! It's so hot here!" said Arlo. Cammy laughed, "I store fat in my hump and I don't need much water!"

Then, Arlo dove into the deep, dark ocean, where he found Glowey the Anglerfish. "It's so dark!" Arlo said. "I make my own light," said Glowey. "It helps me find food!"

Finally, Arlo climbed a tall mountain and met Billy the Mountain Goat. "Careful!" Arlo shouted. "No worries," Billy said. "My hooves are perfect for climbing rocky cliffs!"

After his trip, Arlo smiled. "Animals are amazing! They have special body parts and behaviors to survive anywhere!"

KEY IDEA

Some animals live in extreme environments like icy places, deserts, deep oceans, and high mountains. They survive by having special features like warm fur, strong hooves, glowing bodies, or water-saving tricks!

Activity 87 Where in the World Do They Live?

Help Arlo remember where each animal lives by matching them to their special habitat!

Instructions

1. Draw (✏) a line from each animal to the correct environment.
2. Color (🖍) the animals and habitats.

Polly the polar bear

Cammy the camel

Glowey the anglerfish

Billy the mountain goat

hot desert

deep ocean

rocky mountains

Icy arctic

Build-a-Beast: Survive the Wild! 88 Activity

Instructions

1. Pick one extreme environment: ***Arctic***, ***Desert***, ***Deep Ocean***, ***Mountain***
2. Draw (✏) your own imaginary animal that could survive there!
3. Write (✍) 3 special features that will help your imaginary animal survive.

My choice of environment...

Animal name:

Three (3) special features that will help my animal survive.

Activity 89 Would YOU Survive There?

Instructions

1. Choose one of these animals: Polar Bear, Camel, or Anglerfish.
2. Draw (✏) the animal in its extreme habitat.
3. Write (✍) how it survives using its body parts or behaviors by completing the sentence.

Chosen animal:	Habitat:

This animal survives in the ____________________ because it

has ____________________ to help it ____________________.

Adapt to Survive!

90 Activity

Instructions

With the use of your crayon, color (✎) the correct answer.

1. Camels store fat in their humps to ________________

Stay cool	Keep warm	Store energy

2. Polar bears have thick fur and fat to________________

Help them float	Stay warm	Hide in sand

3. Mountain goats have strong hooves to ________________

Jump high	Climb rocky mountains	Swim underwater

4. Anglerfish live deep in the ocean where it's ________________

Bright	Cold and dark	Hot and sunny

5. Which animal would NOT survive in the desert?

Camel	Penguin	Lizard

Unit 19 Using Models in Science

Activity 91 Mila the Mini-Builder!

Mila was a curious 2nd grader who loved asking questions. One day, she saw a rocket launch on TV and asked, "Can I build a rocket too?"

Her teacher smiled and said, "Let's make a model!" "A model?" Mila blinked. "Like a toy?"

"Kind of!" said her teacher. "A model is a small version of something that helps us see how it works.

Scientists use models all the time!"

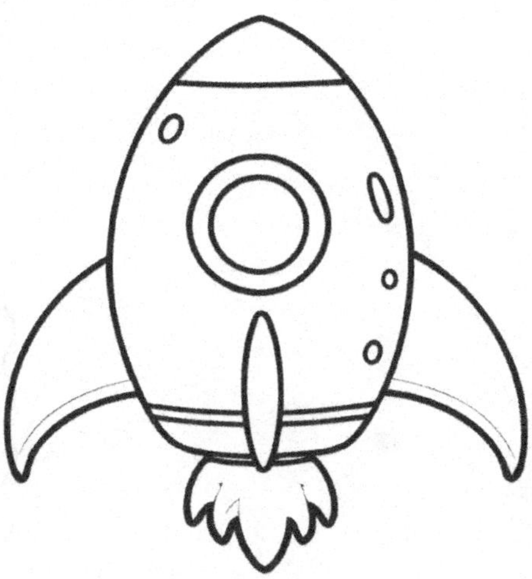

So, Mila grabbed paper, tape, and a straw. She built a mini rocket and launched it across the room. WHOOSH!

"It works!" she cheered.

Later, she made a model of the solar system with foam balls and string.

She even built a model of her teeth using clay during science time!

Each time, Mila learned something new. She learned how planets move, how rockets fly, and how her body works.

She smiled and said, "Models help me understand BIG ideas, one small piece at a time!"

KEY IDEA

Models in science are small, simple versions of big or complex things. They help us understand, test, or show how something works.

Mini World on Paper!

92 Activity

Instructions

1. Choose a habitat: ***forest***, ***ocean***, ***desert***, or ***arctic***.
2. Draw (✏) a model of that habitat including at least one plant and two animals.
3. Label (✍) each part to show where the animals live and what they need to survive.

This model shows a ______________ habitat

where ______________ live.

Activity 93 Who Eats What?

An ecosystem model is a drawing or chart that shows how plants, animals, and their homes all work together. It helps us understand how living things get food and share the same space.

Instructions

1. Draw (✏) a simple food chain using arrows (→) to show how animals get their food.
2. Start with a plant (like grass).
3. Then, draw (✏) an animal that eats plants (called a herbivore), like a rabbit.
4. Next, draw (✏) an animal that eats other animals (called a carnivore), like a fox.
5. Use arrows (→) to show the order.
6. Label (✍) each part of your food chain.
7. Color (✏) your pictures to make your chart bright and fun!

Simple Food Chain

What Does This Model Show? 94 Activity

Instructions

1. Match each model picture to what it represents by drawing (✏) a line between the model and its real-world idea.
2. Color (✏) each model when you're done!

Solar System

water movement

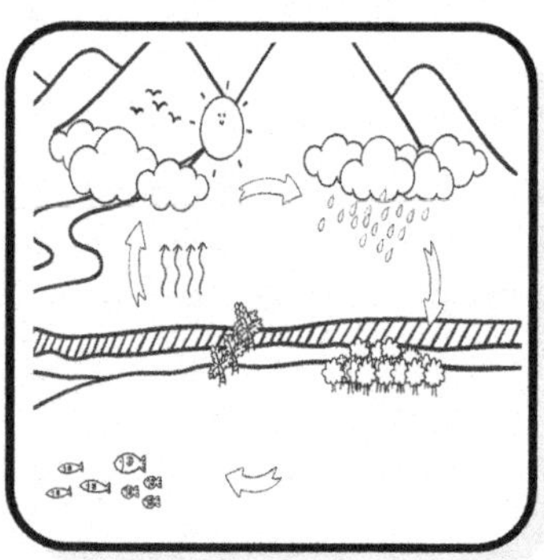

habitat

real house

Activity 95 Model Smarts!

Instructions

Circle (○) the correct answer.

1 A model is:

(a) a real thing

(b) a small version

(c) a picture of food

2 What does a model help us do?

(a) Eat candy

(b) Build furniture

(c) Understand something big or hard to see

3 Which of these is a model?

(a) A toy rocket

(b) A real elephant

(c) A sandwich

4 Which model shows the solar system?

(a) A poster of planets

(b) A beach ball

(c) A rock

5 A diagram that shows how plants and animals live together.

(a) Puzzle

(b) Ecosystem Model

(c) Dance

Unit 20

Review and Reflection

The Science Superstars' Big Mission! 96 Activity

It was the last day of science class, and the students in Room 2A were buzzing with excitement. Their teacher, Miss Luna, stood up and smiled. "Today, you are all becoming... Science Superstars!"

"Science Superstars?" asked Jayden. "What do we have to do?"
"You'll go on a mission to remember all the amazing things you learned this year!" said Miss Luna, handing out shiny science badges.

"We learned about animals and how they survive in wild places," said Maya. "Like polar bears with blubber and camels with humps!" added Leo.

"We also learned about water and where it comes from, like oceans, rivers, and even rain!" said Aaliyah.

"Don't forget how we learned to use models to show things like the solar system or a habitat," said Mateo proudly.

"Oh! And we discovered what floats and sinks, and why!" said Lily, tossing her rubber duck into a bowl of water.

The class laughed and remembered building mini-ecosystems, exploring light and shadows, and learning to protect nature by saving water and energy. Miss Luna clapped her hands. "You remembered it all! That means you're ready for your Science Superstar medals!"

The students cheered, knowing that the science they learned wasn't just in their notebooks but in the way they saw the world now: with wonder, questions, and curiosity.

KEY IDEA

Review and reflection help us remember all the things we've learned. Looking back makes us smarter moving forward!

Activity 97 Find It Like a Scientist!

Instructions

1. Search around your home, school, or outdoors.
2. Find one item or example that relates to each science topic.
3. Draw (✏) and label (✍) each item in the box.

Animal Habitat

Something that floats

Water Source

Living Thing

Memory Match-Up!

98 Activity

Instructions

Review important science vocabulary and concepts by matching them to clear pictures or simple definitions.

evaporation	pulls metal objects
magnet	use something again
recycle	a place where animal lives
roots	water turning into vapor
habitat	gives energy to plants
sunlight	part of the plant that grows underground and absorbs water
rain	water falling from the clouds
thermometer	measures temperature

Activity 99 Are You a Science Superstar?

Instructions

Circle (○) the best answer for each question.

1. Which state of matter is steam?
   - (a) Solid
   - (b) Liquid
   - (c) Gas

2. Which of these is a natural water source?
   - (a) Faucet
   - (b) Lake
   - (c) Water bottle

3. Why do scientists use models?
   - (a) To tell jokes
   - (b) To understand big ideas
   - (c) To win races

4. What causes a shadow to form?
   - (a) Light is blocked
   - (b) It's cold outside
   - (c) Loud sounds

5. What do plants need to grow?
   - (a) Music
   - (b) Water and sunlight
   - (c) Pillows

Activity 100 My Science Discovery Journal

Instructions

1. Think back to your favorite discovery from this workbook and draw (✏) it.
2. Write (✍) about why it is interesting and how you can explore it like a real scientist!

I discovered that

It was interesting because

If I were a scientist, I would explore

Science Superstars Awards! 101 Activity

Instructions

1. Pick your favorite science topic from this workbook.
2. Draw (✏) a picture and write (✍) a sentence about why it was your favorite!

I am a superstar in learning about

favorite Science topic

I liked learning about this because

Science Superstar Certificate

Celebrate your hard work and accomplishments in science!
You get to create and design your very own Science Superstar Certificate!

Instructions

1. Use the template below to create your certificate.
2. Fill in the details and make it colorful and unique!
3. Add drawings or stickers that represent your favorite science topic from the book.

SCIENCE SUPERSTAR CERTIFICATE

This certificate is awarded to

____

name

For completing the 2nd Grade Science Book
and exploring the wonders of science!

________________ ____________________

Date Teacher/ Parent Signature

Unlock Your FREE Science Bonus!

+ Answer Key

SCAN ME

BONUS

- **21 Hands-On Experiments:** Bring science to life with exciting, step-by-step activities you can do at home!
- **Learn Through Play:** Explore core concepts like Matter, Forces, and Energy with fun stories and characters for each experiment.
- **Sparks Curiosity:** A perfect companion to your workbook, designed to make learning interactive, messy, and memorable.

Next Generation Science Standards

Life Science Standards

- **2-LS2-1:** Plan and conduct an investigation to determine if plants need sunlight and water to grow.
- **2-LS4-1:** Make observations of plants and animals to compare the diversity of life in different habitats.
- **2-LS2-2:** Develop a simple model that mimics the function of an animal in dispersing seeds or pollinating plants.

Earth Science Standards

- **2-ESS1-1:** Use information from several sources to provide evidence that Earth events can occur quickly or slowly.
- **2-ESS2-1:** Compare multiple solutions designed to slow or prevent wind or water from changing the shape of the land.
- **2-ESS2-2:** Develop a model to represent the shapes and kinds of land and bodies of water in an area.
- **2-ESS2-3:** Obtain information to identify where water is found on Earth and that it can be solid or liquid.

Physical Science Standards

- **2-PS1-1:** Plan and conduct an investigation to describe and classify different kinds of materials by their observable properties.
- **2-PS1-2:** Analyze data obtained from testing different materials to determine which materials have properties best suited for an intended purpose.
- **2-PS1-3:** Make observations to construct an evidence-based account of how an object made of a small set of pieces can be disassembled and made into a new object.
- **2-PS1-4:** Construct an argument with evidence that some changes caused by heating or cooling can be reversed and some cannot.

Engineering, Technology, and Applications of Science

- **K-2-ETS1-1:** Ask questions, make observations, and gather information to define a simple problem that can be solved through the development of a new or improved object or tool.
- **K-2-ETS1-2:** Develop a simple sketch, drawing, or physical model to illustrate how the shape of an object helps it function as needed to solve a given problem.
- **K-2-ETS1-3:** Analyze data from tests of two objects designed to solve the same problem to compare the strengths and weaknesses of how each performs.

We Appreciate You

Our greatest joy comes from seeing little learners grow, explore, and fall in love with discovering the world around them.

That's why your feedback means so much to us!

Even a few honest words about your experience with our book can make a huge impact. It truly helps more than you know.

We're a small, heart-driven education brand, and every review helps us reach more bright, curious young minds who are eager to learn and dream big.

And yes, it might even help us sell a few more books along the way!

Each review fills us with gratitude and reminds us why we do what we do.

Want to make our day?

SCAN ME

Made in the USA
Monee, IL
14 September 2026

62980094R20063